Version 1.2

First Edition Published 2011, 2020

Technology Template Series: Defect Management Plan - TM 007

Defect Management Plan

Welcome

Welcome to the Technology Template Series and thank you for your purchase.

We trust that you will gain benefit and insight from each of our templates. Our approach is to provide you with a standardised model and access to the knowledge gained over 30 years within the Information Technology industry and offers this in a modularised manner; thus allowing you to purchase only the pieces that you require, without purchasing an entire methodology or text book on the subjects covered.

The structure is to provide you with the layout of each template, with details within each section that define what you are required to write to complete the document and in most cases examples of this.

Purpose and Use

The Defect Management Plan is a deliverable for all projects and programs. The document outlines how Defects will be managed. It also describes who will be involved and the roles and responsibilities of all parties. This e-book provides you with a well-structured template for you to write your own Defect Management Plan, whether for a project or a program

Template Completion

This template is to be completed for all high and medium-sized projects / programs.

You need to complete as much of this template as possible. Where this template is being used for a program, amend headers as appropriate and apply the program context in completing the sections.

All sections in this template are mandatory. However, if there are sections that you do not think are relevant for your project (e.g. Sourcing arrangements), then enter 'N/A' under the section heading with an explanation as to why that particular section is not relevant. Please do not delete any sections from this template.

Additional sections, tables and appendices may be added where appropriate. Existing sections can be altered or expanded as required to obtain stakeholder acceptance or sign-off.

Additional Information

The defect management tool identified in the Defect Management Plan template is Hewlett Packard's *<Defect Management Tool>* (HPQC), but you could substitute this for whatever defect management tool you use within your organisation.

Template

The Defect Management Plan Template follows from here.

<name of Project / Program>

Defect Management Plan

Doc. ID	
Version	
Date	
Author(s)	o
Status	☐ Draft ☐ Under Review ☐ Approved

This template is to be completed for all high and medium-impact projects / programs.

You need to complete as much of this template as possible. Where this template is being used for a Project / Program, amend headers as appropriate and apply the Project / Program context in completing the sections.

All sections in this template are mandatory. However, if there are sections that you do not think are relevant for your project (e.g. Sourcing arrangements), then enter 'N/A' under the section heading with an explanation as to why that particular section is not relevant. Please do not delete any sections from this template.

Additional sections, tables and appendices may be added where appropriate. Existing sections can be altered or expanded as required to obtain stakeholder acceptance or sign-off.

I. Document Purpose

The purpose of this document is to define a high-level plan for defect management including a high-level approach, defect management process, defect management organisation and management controls for a Project / Program and demonstrate how it aligns with the requirements management, release management, testing and quality management capabilities within the program.

The Defect Management Plan is an initial planning document that will provide a framework for the defect management effort and capability uplift for the *<name of project / program>*. This document is intended to provide a controlling framework for defect management for the *<name of project / program>*.

This document is developed during the planning phase of a Project / Program Once approved the Defect Management Plan represents the agreed approach for all teams in all work streams engaged on a Project / Program to follow.

One of the primary roles of this Defect Management Plan is to ensure that there is standardisation of the defect management processes across a Project / Program. That each of the work stream on the Project / Program is engaged to run defect management in a consistent manner and in collaboration across the Project / Program; e.g. all Projects / Programs are using the same terminology, tools, templates, processes, etc. This assists the project / program in managing and tracking the defect management effort and the reporting from each stream is done in a consolidated manner that can easily be rolled up into a program view.

Document Scope

This document relates to defect management for a Project / Program, and associated work streams

Identify and name any work streams and entities covered by the Defect Management Plan here.

The Defect Management Plan is intended to also accommodate future additional work streams; it sets the standard for future projects / programs for *<name of company>*.

Document Audience

This document is a key project / program deliverable. It is issued to all parties that have been assigned responsibilities in this Defect Management Plan. This includes, but is not limited to for example:

- *Project / Program Manager*
- *Technology Project Managers*
- *Business Project Managers*
- *The Project Management Office*
- *Test Managers*

- *Test Teams*

- *Systems Development*

- *Architecture Team*

- *Business Users, including Subject Matter Experts (SMEs)*

- *Risk and Issue Management*

- *Release Management*

- *Delivery Team Representatives*

- *Test Environment Management Team*

- *All staff with a RACI role for this document*

Related Documents

Provide a list of all related and referenced documents including Title, Version and Link or Location of the source document

Table 1 - Related Documents

Document Title	Version	Link / Location

Change History

Provide a chronological history of changes to this document; include version number, date of change, name of who made the changes, and a brief outline of the changes made.

Table 2- Change History

Version	Change Date	Name	Brief outline of changes

Program Contact List

Provide a list of relevant contacts for the Project / Program; identify Role, Name and Contact details.

Table 3 - Project / Program Contact List

Role	Name	Contact Details
Program Sponsor		
Business Program Director		
Technology Program Director		
Test Manager		

Defect Management Plan Sign-off

Identify the approvers of this document; list type, name, title, date. Allow for signature to be included in table, this can be either formal written signature or attach electronic approval, e.g. e-mail.

Table 4 - Defect Management Plan Sign-off

Type	Name	Title	Signature/Artefact	Date
Approvers				

Review of the Defect Management Plan

Identify the reviewers of this document. List name, title, date. Allow for signature to be included in table, this can be either formal written signature or attach electronic artefact, e.g. e-mail. Refer also to a reviewer feedback form that you should use for all documents to be reviewed, this gives you a standardised format for review comments. These can then be inserted as objects within this document.

Table 5- Reviewer List

Reviewer Name	Title	Date	Artefact

Introduction

Executive Summary

Write up the executive summary of the Defect Management Plan here.

Program overview

Document a high level description of the Project / Program here.

Noting that the objective of this document is to provide a broad outline of the <name of Project / Program> defect management plan.

Background

Write up a short background for the drivers for initiation of the Project / Program and the need for a defect management plan. Include any reference and technology strategies applicable.

Defect Management

Scope

Identify the scope of your defect management plan, for example:

Broadly speaking, defects will originate from either Quality Assurance (peer review) or testing activities (static testing during test planning, and dynamic testing during test execution).

However, "defects" identified from peer reviews are normally documented as part of the Project / Program Document Management process via Review Feedback Sheets and not as defects per se.

Hence the scope of defect management will provide exclusively for:

- *Defects raised during test planning, preparation and execution for all of the Test Phases (e.g. Unit Test, System Test, SIT, UAT, etc) and applications under test, as defined in the Master Test Strategy and Master Test Plans. Such defects mainly relate to code and/or requirement issues. In addition, defects raised during testing may relate to data, test environments, release deployments, etc.*

Defect management will not cater for:

1. *Defects (or "proto-defects") raised in document peer reviews, prior to the first requirements baseline. These are managed via the Project / Program's Document Management process.*

2. *Data Quality Issues in Production. These are logged and managed via a separate <Defect Management Tool / Help Desk Tracking Tool> (and can be referenced during <name*

of project / program> testing by using the foreign defect id if necessary).

Objectives

List the objectives of your defect management plan. For example:

The objectives of the Defect Management Plan are to:

- *Define defect recording standards; including mandatory and non-mandatory fields, together with standard values for fields within the <Defect Management Tool>*

- *Document realistic defect turnaround times (service level targets)*

- *Enable Testing metrics based on defects, including both analytical (e.g. root cause analysis) and predictive metrics (e.g. defect closure rate versus testing timeline)*

- *Provide a reliable, transparent, auditable and pro-active defect management process for all current and future projects, including defect triages (to ensure that the right defects are dealt with initially) and defect metrics (to provide relevant and timely information to facilitate decision making during testing).*

- *Define a defect management process and process workflow that aligns with the testing and release management (Software Configuration Management) processes used on the Project / Program*

- *Explain defect management tool usage and integration along with any automated work flows*

- *Define roles and responsibilities, including organisation of defect management activities*

Assumptions & Constraints

List the major assumptions made in formulating this Defect Management Plan. State any major constraints.

As Assumptions & Constraints are identified, they should be recorded in a central Assumptions & Constraints register located in the Project / Program folder.

Assumptions & Constraints are grouped into the following categories, and are numbered sequentially within each section.

As Assumptions & Constraints are identified, the Process for validating the assumption or constraint is identified, and a review date is determined. Once an assumption or Constraint is validated (or rejected), the resolution is updated in the Action/Outcome column.

The Assumptions & Constraints register is to be reviewed regularly throughout the rest of the Project / Program. It is expected that at the conclusion of the Program, all Assumptions & Constraints will have been actioned. example categories:

Table 6- Assumption Register Groupings

Number	Group Description
1	Cost/Financial
2	Time/Schedule
3	Technology
4	Resource
5	Business Requirements
6	Governance
7	Engagement
8	Quality Requirements (NFRs)
9	Program Quality
10	Information Requirements
11	Approach
12	Regulatory and Statutory Compliance
13	Market Window

Defect Management Process

Workflow

Identify and document the workflow for your defect management process here, for example:

The Defect Management Process workflow is shown in the diagram below (start at 1):

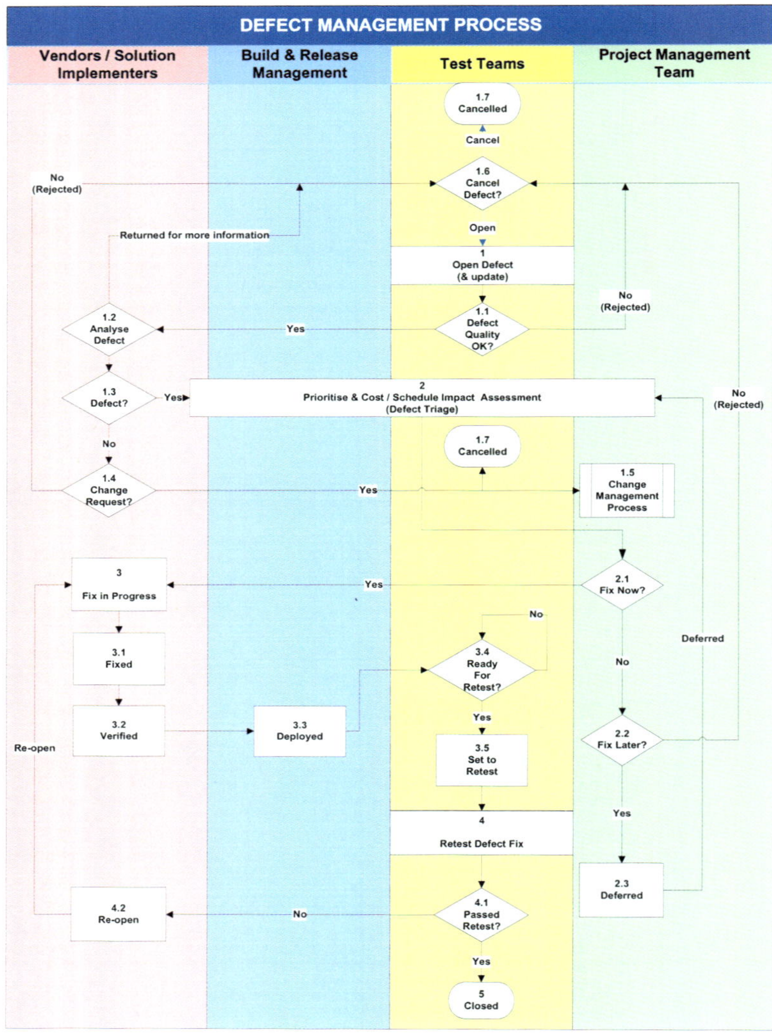

Figure 1- Defect Management Workflow

Defect Definition

IEEE defines a defect as a product *anomaly*, and defines an anomaly as:

"Anything observed in the documentation or operation of software that deviates from expectations based on previously verified software products or reference documents."

Defects include but are not limited to any of the following:

- fault
- failure
- degradation
- deficiency
- error
- non-conformance to specification
- inconsistent, ambiguous, incomplete or otherwise incorrect specification
- missing functionality (function is specified but not delivered)
- unspecified functionality (function is delivered but not specified)
- functionality performance below or not in accordance with specification, or partial or non- performance
- unresolved data cleansing issue that impacts testing or would impact business operation in production
- unresolved test environment issue that impacts testing

During testing, a defect will be raised in any of the above instances. Alternatively an issue or change may be raised when the possible outcome of testing or implementing the integrated component set as currently defined, could result in the delivery of a sub-standard solution.

All defects are to be recorded in *<Defect Management Tool>* immediately by the testers, with test leads reviewing progress regularly.

Defect Priority / Severity Definitions

Defects will have two classifications regarding fix urgency; priority and severity.   An example of a definition of these classifications is depicted in figure 2 below.

Your organisation may have alternate definitions, adjust your plan accordingly.

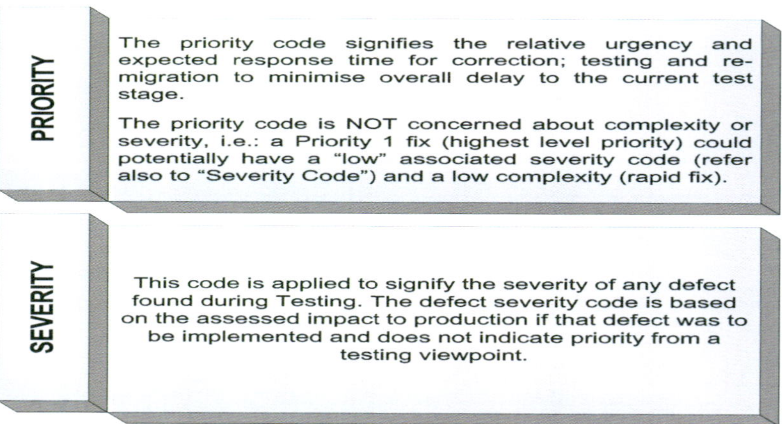

Figure 2- Defect Priority / Severity Definitions

Note that defect severity and priority ratings for outstanding defects are usually used as *exit criteria* from a test phase. Refer to your project / program Master Test Strategy and individual Master Test Plans (per work stream) for specific details.

In addition production deployment of an application will usually not be allowed in most organisations, if any severity 1 or 2 defects are outstanding.

Defect Priority

Table 7- Defect Priority

Defect Priority Levels	
1-Urgent	Prevents progress of the testing program (show stopper) Most test cases are impacted
2-High	Severely limits the progress of the testing program Many test cases are impacted
3-Medium	Limits the ability to perform some tests Some test cases are impacted
4-Low	Has minimal or no impact on the ability to continue with testing At worst only a few test cases are impacted

Defect Severity

Table 8- Defect Severity

	Defect Severity Levels
1-Critical	This is a critical defect that would create severe disruption to the business operation or stop a significant portion of the business operation. For example: • All users would be unable to use an application. • Key business functionality compromised and no workaround exists
2-Major	Significant impact on business operations and no acceptable workaround exists. For example: • A failure in one application affecting other downstream systems. • A major application problem causing considerable system downtime, financial penalty or loss of integrity with customers. • Ability to meet target customer levels is partially compromised
3-Moderate	Business operations are impacted but an acceptable workaround exists.
4-Minor	Minor system impact. An acceptable workaround exists.

Severity and Priority Combinations

Table 9- Severity and Priority Combinations

	Severity 1	Severity 2	Severity 3	Severity 4
Priority 1				
Priority 2				
Priority 3				
Priority 4				

In the absence of any further guidance, defects should be addressed in the following order:

Within this broad framework, high priority defects can require attention before equally high severity defects as the impact on the testing schedule *now* can override concerns over potential business impact *later.* In the end, however, both high priority and high severity defects require urgent resolution.

This approach is then refined through the Defect Triage process. For example, it might make sense to fix some low priority defects when fixing a high severity defect as all the fixes are in the same code module and so the module (and defect fixes) only need retesting once.

Service Level Targets

All defects are to be resolved using best efforts. This implies the remediation of major and critical defects as a priority and minor defects in the shortest timeframe practicable. Defect Service Level Targets apply to Vendor / Solution Implementer Teams. Defect Service Level Targets should be agreed, sensible, manageable and achievable. However, initial measurable targets are needed. Hence, the following *1-2-4-8 Service Level Targets* are suggested:

Table 10 - 1-2-4-8 Service Level Targets

Severity or Priority (Highest Value)	Action Required	OPEN defect assigned to Vendors / Solution Implementers for ANALYSIS or FIX. Initial response / update required within:	Service Level Target For Fix (In Working Days)
1	Immediately, within current test cycle	1 Hour	Hot fix to be deployed, or bundled in next deployment, **within 1 day**
2	As soon possible, within current test cycle	2 Hours	Hot fix may be deployed, or bundled in next deployment, **within 2 days**
3	As soon possible, before end of next test cycle	4 Hours	Bundled in earliest available release, **within 4 days**
4	As soon possible, before end of next test cycle	8 Hours	Bundled in earliest available release, **within 8 days**

Combining Table 7 and Table 8, you get:

Table 11- Defect Fix Service Level Targets by Severity and Priority Combinations

	Severity 1	Severity 2	Severity 3	Severity 4
Priority 1	1 day	1 day	1 day	1 day
Priority 2	1 day	2 days	2 days	2 days
Priority 3	1 day	2 days	4 days	4 days
Priority 4	1 day	2 days	4 days	8 days

Sign-off by all parties of this Defect Management Plan, constitutes agreement with these proposed Service Level Targets as *<name of project / program>*'s *initial* measurable targets. Note that vendors / solution implementers may propose workarounds that help alleviate issues blocking test execution, though such workarounds do not change measurement of actual fix times against Service Level Targets.

Note that a Service Level Target is similar to a Service Level Agreement (SLA). However, SLAs are often contractually binding whereas Service Level Targets are used internally on a Project / Program to provide defect metrics, in order to highlight any issues with defect turnaround times.

These Service Level Targets do not supersede any defect turnaround SLAs (contained within contracts or SoWs) in any way. Note that managing defect turnaround times against contractually binding SLAs is usually the responsibility of the work stream Project Managers.

Defect process – working with the Vendor(s)

Wherever possible, Vendors / Support teams for the *<name of project / program>* should use *<name of project / program>*'s *<Defect Management Tool>* as the repository for managing testing defects. If the vendor does use some other tool for managing defects and does not wish to use *<name of project / program>*'s tool set – *<Defect Management Tool>*; then a mechanism for linking the two must be established. The defect id in the vendor's tool must be passed back and added to the *<name of project / program>* defect record (and vice versa). Therefore capability via an enterable field within the defect record must be available to identify "foreign" defect ids. Refer to flow chart below for an example.

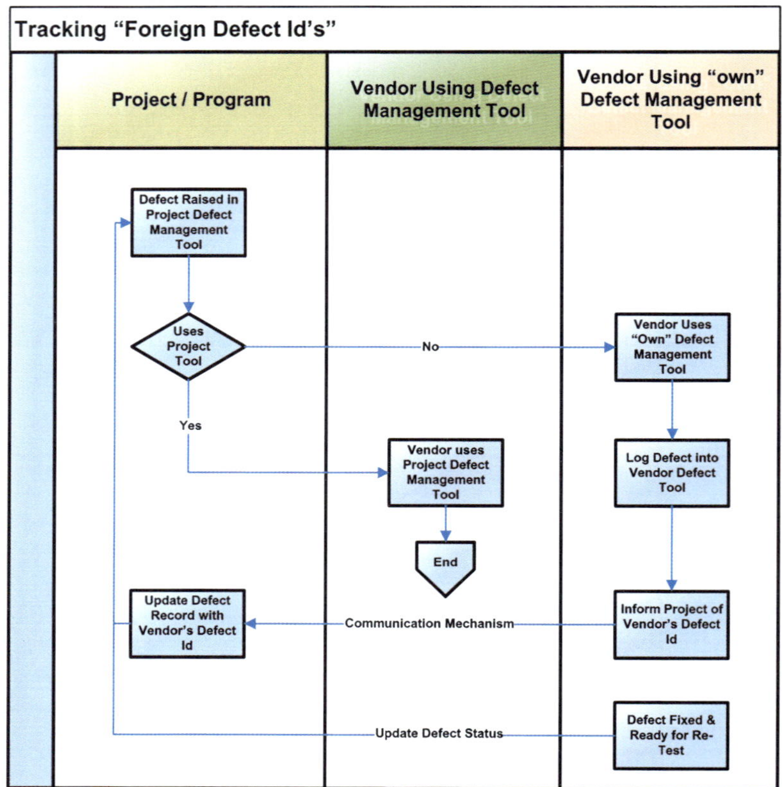

Figure 3- Foreign Defect Id Workflow

Note that this process can also be used to link to Production Data Quality Issues managed by via that *<Defect Management Tool>*.

Fields

Note that this does not include any standard *<Defect Management Tool>* fields such as Detected By, or Assigned To. In addition, *<name of company>* standard fields and values are assumed unless specified below.

Table 12- <Defect Management Tool> defect fields

Field	Value	Description	Notes
Status			Mandatory
	New	Default defect status for newly raised defect. Test analyst changes to Open when assigning to Test Manager for review.	
	Open	With test manager / test team for review / analysis of defect details.	
	Analysing	Under review and analysis by member of solution team (developer, business analyst etc) to confirm that it is a defect and to establish the required fix, or establish that this is in fact a CR.	
	Fix In Progress	Required fix is known, assigned to member of solution team to fix.	
	Fixed	Fixed by member of solution team, and assigned to Release Manager to incorporate into a release.	Also see **Cause** field
	Verified	Deployment verified by member of solution team in own development / test environment.	
	Deployed	Successfully deployed to test environment and ready for re-test. Assigned to test manager (to assign defects).	
	Re-test	Newly deployed defect fix now assigned to test analyst for re-test.	
	Re-open	Has failed re-test, assigned back to solution team to re-assess and re-fix.	
	Deferred	Defect is not to be fixed for now, but may be re-considered for fixing in a later release.	
	Rejected	Defect is unable to be reproduced, or lacks sufficient details – returned to test team to update (and then mark Open) or cancel.	
	Cancelled	Defect did not require fixing or re-testing, and after further consideration is cancelled.	This is a valid end status for a defect. Also see **Cancel Reason**
	Closed	Defect is successfully fixed and tested.	This is a valid end status for a defect.
Field	**Value**	**Description**	**Notes**
Cancel Reason			Mandatory when defect **status** changed to Cancelled

Automated *<Defect Management Tool>* defect workflows

Document the defect workflows you require for your Project / Program. You will find that this will evolve over time. Ideas include:

- *Auto email on change of Assigned To*

- *Assign team field based on Assigned To value*

- *If defect closed then auto email Detected By person*

- *Auto email any New Priority 1 and Severity 1 (i.e. both severity and priority are 1) defect to Test Manager, and/or Defect Manager*

- *Select project and application list is then pre-populated with only applications relevant for that project*

- *Select application and release list is then pre-populated with relevant release numbers (if this is not possible then release number should be a free text field that becomes mandatory when application is selected)*

- *If defect Status marked fixed then release number becomes mandatory*

- *If defect status marked fixed then Cause field becomes mandatory and pre-populated with relevant values (Existing, requirements, code, data, environment, testing, or multiple)*

- *If defect is cancelled then defect Cancel Reason becomes mandatory and Cancel Reason list pre-populated with relevant values (with Change Request, duplicate, raised in error etc)*

- *If defect is set to Open or Re-Open then Cancel Reason - if present - is blanked*

- *If defect Status marked Rejected then Assigned To pre-populated with Detected By field value*

- *Test Phase is pre-populated with value from test case if raised during test execution*

- *Defect state transition – based on Status – can be limited / enforced via work flow (e.g. prevent users from marking an Open defect Closed as a defect must be in Re-Test Passed in order to be Closed)*

Related Methodologies

Identify any related approaches/methodologies that your Project / Program will be using.

The <name of project / program> will use the following methodologies that relate to defect management:

- *Software testing activity during test planning and test execution will generate defects. See also Master Test Strategy. However, the methodology that binds Software Testing and Defect Management together is Release Management.*

- *Release Management including software configuration management. See also Release Management Plan.*

- *Requirements Management. Requirements affected by peer review "proto-defects" and software testing defects, plus CRs. See also Requirements Management Plan.*

Software Testing

The purpose of running a test case can be viewed as checking the theory that the code matches the specification (and then providing the critical information on the result – pass or fail).

Yet for each test condition in a test case, there are many ways for that test case to fail but only one *valid* way for a test case to pass. The following truth table - simplified when compared to the often highly complex reality of software testing - attempts to demonstrate why:

Table 13- Truth Table For A Test Condition

Test Case	Code	Specification	Result
OK	OK	OK	Pass
OK	OK	Not OK	Fail
OK	Not OK	OK	Fail
OK	Not OK	Not OK	Fail
Not OK	OK	OK	Fail
Not OK	OK	Not OK	Fail
Not OK	Not OK	OK	Fail
Not OK	Not OK	Not OK	Fail

If the specification is correct, the code matches the specification and the test case matches the specification for the specific test condition then when the test case is run it *should* pass. Otherwise, the test case *should* fail, resulting in a defect being raised. This can make it very challenging to correctly pass a test case.

Therefore it is critical – for *each* CR and *each* defect in a release – that versions of the specification and the code are aligned when delivered to testing, and that the test case is also aligned (i.e. updated if necessary). Release Management helps to achieve this by ensuring complete transparency and traceability of the contents of each release, including every change item (CRs and defects) and every configuration item affected (code or requirement). It also attempts to ensure that, when audited, the history of every CI is also transparent and therefore traceable back to the change items (CRs and / or defects) that impacted them.

However Release Management errors (e.g. the wrong version of a release note given to the test team), test execution errors (e.g. test case, code and specification all match but wrong the test case actually run or the right test case is run but done so incompetently), test data issues (e.g. manufactured test data lacks data integrity) and test environment issues (e.g. unlicensed or incorrectly installed software) can make passing a test case even harder. Or, allowing for false positives, such factors can also make it easier to incorrectly pass a test case. If such factors were included in the truth table above then the number of possible fail combinations would increase exponentially.

Release Management

During software testing Release Management is the "glue" that holds change management and defect management together. Hence, Release Management is critical to the success of a project or program. Without Release Management software testing can become increasingly inefficient or at worse descend into chaos. Software testing is highly sensitive to release management as test cases verify requirements against code for each release. Hence, requirements and code must be aligned.

Release Management is a methodology for ensuring controlled drops of aligned code and requirements documentation into test environments. The following diagram shows the relationship between Release Management and both Change Management and Defect Management:

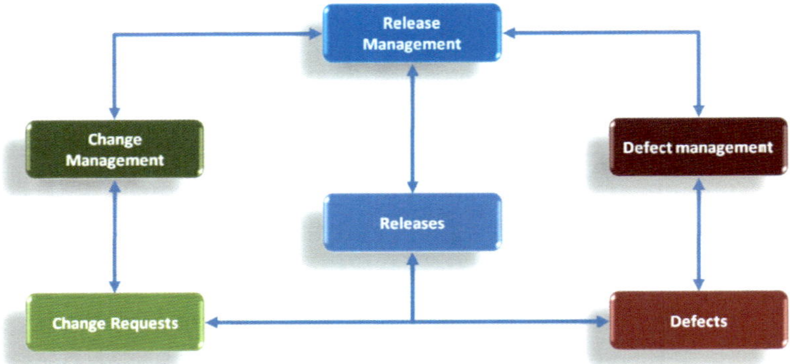

Figure 4- Release Management, Change Management & Defect Management

The diagram also shows that change management involves multiple CRs, release management involves multiple releases and defect management involves multiple defects.

Note that Releases are tied to calendar dates with consideration for deployment moratoriums, application and infrastructure inter-dependencies, parallel releases for a specific application (see Figure 7), and solution delivery capacity etc. to be outlined in the Release Management Plan. A single Release then is comprised of one or more change requests and / or one or more defects (the exact mix of CRs and defects will vary from release to release) together with a *mandatory* release note:

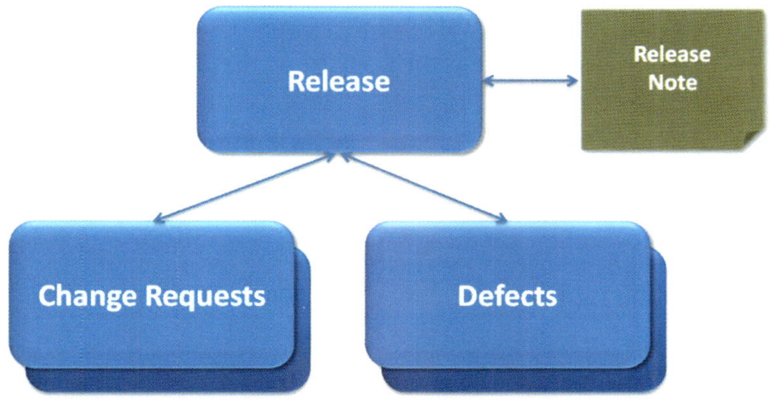

Figure 5 - A Single Release

It is critical to note that that the concept of Releases only applies to configuration items for a software product, meaning the code and all related requirements specifications:

Figure 6- Software Configuration Management – CI Lists and Repositories

A program such as *<name of project / program>* may have multiple Product Configuration Item Repositories (i.e. at least one per project). With *<name of project / program>* tools such as ReqPro are a common physical Product Configuration Item Repository for requirements, though logical separation will occur for each unique work stream product. In addition, each work stream will have one or more Product Configuration Item Repositories for their code (such as Clearcase).

Non-Product deliverables may still be put under version control (or document management) but, from a testing perspective, changes to these deliverables are not subject to defect management, change management or release management.

In addition, for any given application under test, there can be more than one release being built or being tested. In this simple example Release 1.0 is in Production, Release 1.1 is in UAT and Release 1.2 is under System Test in a different test environment. The possibility exists of an emergency fix to Production that would then need to be applied to Release 1.1 and Release 1.2.

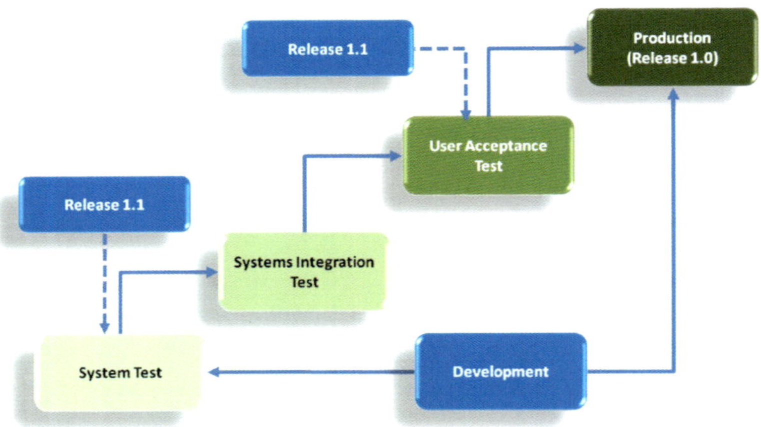

Figure 7- Multiple Releases For a Single Application (example only)

Baselines and Defects

From a defect management perspective, it is important to understand when it is acceptable to raise defects against both requirements and code.

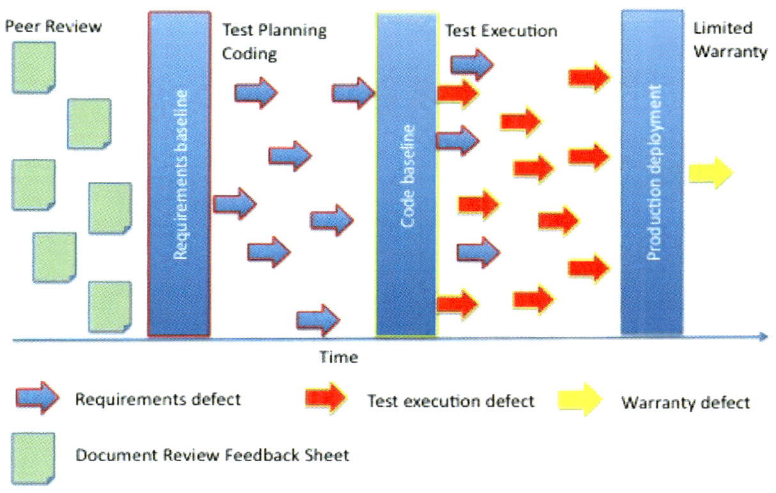

Figure 8- When can a defect be raised?

The Document Management process applies to all *<name of project / program>* documents (product and non-product). This involves QA Peer Review activity whereby a Document Review Feedback Sheet (one per reviewer) is used to capture document feedback. Once the document is updated and approved it is baselined and formally released for both coding and test planning to commence. After this point the Defect Management process applies, and requirements defects discovered during test planning and test execution are raised in *<Defect Management Tool>*.

Later the code is also baselined and formally released to the test team (and code is deployed to the test environment). At this point test cases are executed and code-related defects can also be raised in *<Defect Management Tool>*.

During the limited product warranty period defects may be still be raised in *<Defect Management Tool>*, after which the BAU production incident process applies.

Baselines for requirements and/or code are to be agreed by the business and reviewed by the Release Manager.

Major baselines are typically given major release labels (e.g. Release 1.0, or Release 2.0), with minor releases causing the minor release number to increment (e.g. Release 1.1, or Release 2.1). The exact release numbering convention, and how that relates to documentation or code baselines, which is outlined in your Release Management Plan.

Defects and Release Management - Responsibilities

The following Release Management responsibilities apply to defects:

- If fixing a defect always use the defect id every time when checking in your fix (for documentation and code). Note that the same defect id may apply to multiple configuration items. Also, multiple defect fixes may be applied to one configuration item and so each relevant defect id should be included in the check-in comments.

- For documents, always update the document history with the defect id(s) of any defects that you have fixed in that document. However, note that automated updates of defect ids against requirements (e.g. from *<Defect Management Tool>* to ReqPro) will occur once *<name of project / program>* Tool Enablement has been implemented.

- The Release Manager is responsible for periodically auditing check-in comments, and document histories.

- The Release Manager is responsible for ensuring the accuracy of the Release Notes, including defect details (defect id, defect summary, test deployment details) of defects fixed in this release. It is critical that the Release Manager ensures that only the agreed defects (as per defect triage process) are fixed in each release. The Release Manager should highlight any exceptions as early as possible.

- The Test Manager is responsible for ensuring that Release Notes are validated against expectations and previous releases. Issues with Release Notes are taken up with the Release Manager.

- Once Release Notes have been validated the Test Manager is responsible for approving the test deployment of the release. Once the release is deployed to the test environment then the Test Manager is responsible for changing defect status from Fixed to Re-test and changing *'Assigned To'* to the relevant test analyst for re-testing.

Similar release management responsibilities also apply to CRs – refer Requirements Management Plan, and Release Management Plan.

QA Plan and "Defects"

As noted in the Quality Assurance Plan, the primary focus of quality assurance is early detection and removal of defects through planned peer reviews (a formal type of static testing). This entails a use of the term *defect* in its broadest sense – a product anomaly (see Definitions). Defects (in the broadest sense) are a critical means of assessing product quality based on planned QA (peer review) and testing (test planning and test execution) activities.

However, prior to requirements being baselined the outcome of peer reviews will be Document Review Feedback Sheets (a type of Quality Record), not actual defects. Once requirements are baselined then the outcome of any further static testing (e.g. during test planning) or dynamic testing (during test execution) will be formal software testing defects raised in *<Defect Management Tool>*.

From a Defect Management perspective it is important to distinguish between these two types of defects. Hence peer review "defects" will be referred to in this document as "proto-defects" (a deliberately distinctive term). These early types of defects from peer review activity are managed to resolution via the Document Management process, not the Defect Management process.

Refer Quality Assurance Plan and Document Management Standard & Procedure for more complete details.

Defect Management Approach

The Defect Management Plan is an initial planning document that provides a framework for the defect management activities to be conducted by the *<name of project / program>* work streams.

The Defect Management Plan is intended to provide a controlling framework for defect management for all applications across all work streams participating in the *<name of project / program>*. The *<name of project / program>* Defect Management Plan is a deliverable of the Program.

Raising "Defects" during Peer Review

The Document Review Feedback Sheet is a Peer Review tool used for capturing multiple "proto-defects", with one sheet per reviewer. These "proto-defects" will be managed to resolution via the Project / Program's Document Management process, not the Defect Management Process. These "proto-defects" are *not* to be recorded in *<Defect Management Tool>*.

Once a document under review has been updated and approved then it is made available for coding to commence and for test planning. A requirements baseline (as noted earlier) is the logical demarcation point between the existing Document Management process and the Defect Management process.

Raising Defects during Test Planning

Once requirements are baselined and released for coding and test planning then it is acceptable for a test analyst and/or test team leader to raise a defect against requirements in *<Defect Management Tool>* due to static testing activity during test planning. The Defect Management process and the Release Management process both apply to any such defects.

Note that the Defect Management process workflow ensures that defects are reviewed to confirm whether or not they are true defects or Change Requests (CRs).

Raising Defects during Unit Testing

Once coding starts, developers are encouraged to keep track of any unit testing defects in their own way, and are not *required* to formally record such defects in *<Defect Management Tool>* (though they are encouraged to do so). Defects that are not logged in *<Defect Management Tool>* - informal defects - are not managed via the Defect Management process.

However, if known problems or issues exist at the point of releasing code to test then these should be documented by the development team and then reviewed by the test team to consider raising formal defects in *<Defect Management Tool>* on behalf of the development team. Such defects *are* managed via the Defect Management process.

Raising Defects during Test Execution

Test execution can commence once baselined code is deployed to the test environment.

A test execution defect, when detected and entered into *<Defect Management Tool>*, should initially be analysed by the test analyst and/or test team leader to determine if the defect is related to hardware, data configuration, software, design, and documentation or other. Subject matter experts from the work streams may be used to refine the definition of the defect, its severity/priority and its impact.

Significant defects, which may impact on other work streams, will be entered into *<Defect Management Tool>* by the work stream Test Manager (or delegate) to allow for communication to others. The resolution of these defects will be co-ordinated by the Test Manager; however the work stream Test Manager continues to own those defects and must ensure they are appropriately tracked in *<Defect Management Tool>*.

During testing, defects identified by testers will be entered by testers or their team leads, into *<Defect Management Tool>* to facilitate reporting and tracking. All defects will be reviewed by the Test Manager or Test Team Lead to ensure they are appropriately defined and categorised. The Test Manager will co-ordinate the resolution of these defects and reports of these in the most appropriate manner.

The Test Manager must be fully supported by the work streams and similar 'tier 3' support groups to ensure management to completion of any defects which arise.

Any changes to the configuration or design which are the result of resolution of a defect, MUST be agreed by the owner of the component, recorded using software configuration management processes, and where necessary, reflected in the relevant design or similar documentation.

Note that defects raised during test execution (running test cases in the Manual Runner HPQC window) will be automatically linked with the requirement(s) previously associated with that test case (when building test coverage). However, for defects raised independently of the Manual Runner HPQC window the tester will need to manually link the defect with the associated requirement(s).

Defect Management Principles

- Defects are a critical project communication medium so maintain defect accuracy
- Try to keep to one "fix" per defect, but be pragmatic
- Defect language must always be neutral and factual
- The defect summary should be precise, concise and understandable at a glance
- Use comments to maintain a log of key findings and decisions
- Requirements defects must always include a requirements reference
- Test execution defects must include steps to reproduce (and test data used)
- Test execution defects must include evidence of failure
- The person that raised the defect should be the person that re-tests the defect when it is fixed
- The only final statuses for a defect are *Closed* or *Cancelled*. *Duplicate* is a reason to cancel a defect. *Rejected* is a temporary status only, and must be agreed

by the developer and test analyst involved before be marking *Fixed* (then *Closed*) or *Cancelled*. *Deferred* is also a temporary status that must eventually be *Fixed* (and *Closed*) or *Cancelled*.

- Reasonable Service Level Targets are in place for defect fix turnaround times, and are met. Contractual SLAs may also apply.
- Defect management is transparent, auditable and measurable
- Defect Management must cater for internal and external parties (e.g. vendors)
- Avoid defect ping-pong. If you notice that a defect has been passed back and forth repeatedly without ownership and progress towards resolution then call a meeting to resolve the situation and prevent further waste of time.
- The earliest that defects can be raised is after a configuration item (code or document) is part of a formally baselined release
- The latest that defects can be raised is during limited warranty (after which the BAU Production Incident process applies)

Defect Triages

As the number of defects increases so does the difficulty in ensuring that the right defects are receiving the appropriate level of attention at the right time. The purpose of a defect triage then is to address this problem. This concept comes from the medical profession, whereby patients are prioritised based upon the urgency of their medical condition.

Process Workflow

Figure 9- Defect Triage Workflow

Guidelines

- Although defect triages typically focus on newly opened defects they can be just as useful for reviewing a backlog of older defects.
- The owner of a defect (based on the *Assigned To* field) is responsible for ensuring that defects are up to date prior to each scheduled Defect Triage (or that up-to-date information is available at a defect triage and that the defect will be updated immediately following the defect triage).
- The Defect Manager is responsible for ensuring agreement has been reached with the BPM for any defect where Change = Yes.
- Typically defects will be reviewed in descending priority (*actual* testing impact now) order, followed by descending severity (*potential* business impact in production) order. However, depending upon need, other fields may also be used. For example, sorting on *Application* for a defect fix push to stabilise a particular component or function, or *Target Release* to lock in defect fixes for the next two releases or by a developer user id in the *Assigned To* field (before the developer goes on holiday).
- The chairperson of a project defect triage is the project Defect Manager (typically the project Test Manager or delegate).
- The chairperson of a program defect triage is the program Defect Manager or delegate.
- The Defect Manager sets the pace of the defect triage.
- The Defect Manager organises meeting invites and any conference calls required.

- Meeting invites should allow for time zone differences (e.g. Bangalore and Melbourne).
- The Defect Manager is responsible for preparing and distributing the list of "unresolved" defects to be triaged. Minimum detail (extracted from *<Defect Management Tool>*) includes:
    - Defect Id
    - Summary
    - Project (custom field)
    - Application (custom field)
    - Status
    - Change
    - Priority (desc order – 1st sort field)
    - Severity (desc order – 2nd sort field)
    - Detected By
    - Date Raised
    - Assigned To
    - Foreign Defect Id (if applicable)
    - Target Release
- "Unresolved" or "outstanding" defects include any defect not *Closed* and not *Cancelled*. *Fixed*, *Deployed* and *Re-test* defects can usually be skipped over.
- Where possible *<Defect Management Tool>* should be available and projected on a screen or wall (with the view matching the details above). Live updates can then be made to defects.
- Verbal explanations from involved parties (internal and external) must be concise and precise.
- Resolve sticking points* offline, prior to the next defect triage.
- Defect triages may be scheduled (sometimes daily, but at least weekly) or ad-hoc, depending upon need.

- A defect triage may be conducted at the project level or program level based on need, as agreed by project / program management.
- The maximum duration of a defect triage should be one hour, but an average of 30 minutes is encouraged.
- It is possible that low priority / severity defects will not get discussed, so the defect manager is encouraged to note the age of listed defects and thus "promote" older defects into the discussion.
- A typical defect triage will include representatives from development, business analysis, testing, release management, change management, business SMEs and project management. The absolute maximum number of people involved (including the defect manager) should be 10, but 5 is strongly recommended as a practical average.
- Defect Triages need to take account of Change Management activities, and hence should consider the impact of CRs on each release

Where the cause of a defect is in dispute by one or more parties, specialist assistance (e.g. architecture) may be requested. Alternatively, a resolution may need to be determined 'committee style' by representatives of the various solution components in question.

Outcomes

Outcomes from Defect Triage activity can be broadly categorised as either focussed and immediate or general and long-term.

The immediate outcomes of a defect triage should be that, if only for a select sub-set of the program or project's defects in a specific defect triage, the involved parties are clear which defects are being addressed and why. They should also be in agreement that the right defects are being addressed first. In short, the defect triage provides focus and cohesion on the short-term goals.

The general and long-term outcome of repeated defect triage activity is that an expectation is set that defects must be accurate and up-to-date at all times if the team is to function as efficiently as possible. Everyone involved in managing, analysing, fixing, re-testing or deploying defects increases their awareness of each other's roles and strives to avoid the costly inefficiencies of poor defect quality and of poor defect management. Planning for releases is much more proactive than reactive.

Defect Management Controls

The defect management process itself must be transparent, auditable and measurable. This applies both at the work stream (project) level and at the program level. Additional controls may be enabled through automated *<Defect Management Tool>* defect workflows (see *2.3 Automated <Defect Management Tool> Defect Workflows*). Most controls will rely on Defect Metrics (see *6.1 Defect Metrics*).

Work stream Controls

At the work stream level the Test Managers (or their delegated representatives) will typically fulfil the Defect Manager role for their individual work streams, and ensure that defect quality and defect management effectiveness are proactively monitored and managed. To ensure that defect fix turnaround timeframes are met the Test Managers will actively monitor defect age and closure rates against signed-off Service Level Targets. Preliminary root cause analysis is also the responsibility of Test Managers.

Where work stream defect management metrics may highlight risks / issues such as missed testing milestones, defect turnaround bottlenecks or test environment deficiencies then such risks / issues will be raised and managed according to Program Risk and Issue Management processes.

Program Controls

The Defect Manager may sit in on or chair work stream level defect triages as required, and will chair program defect triages. At the Program level the Defect Manager will monitor and audit defect management activity and ensure consistency across work streams. This will include review of work stream defect metrics, test status reports, test summary reports and QA reports (for peer review defects). The Defect Manager will conduct regular spot checks of defect quality in *<Defect Management Tool>* and defect granularity across work streams, and raise issues with the individuals concerned or the relevant Test Manager. If not resolved then a formal issue will be raised as per the Project / Program Issue Management process.

The Defect Manager will regularly review and discuss any root cause analysis performed by the work stream test managers, and will also conduct a program level root cause analysis in order to highlight any broad trends.

As noted earlier, the Release Manager is responsible for auditing that check-in comments and document histories use defect ids.

To ensure that defect management controls are effective and comprehensive, *<Defect Management Tool>* will be configured to ensure that no defect can ever be deleted from the *<name of*

project / program> project database. Hence, for example, defects raised in error and duplicate defects will be retained in the *<Defect Management Tool> <name of project / program>* database in order to facilitate later analysis of defect data.

Defect Management Tools

Identify what defect management tools that the *<name of project / program>* will use; the following is a recommended tool set for *<name of company>*:

Table 14- Tool Usage For Defect Management

Tool	Purpose
Hewlett Packard Quality Center *<Defect Management Tool>* (*<Defect Management Tool>*)	Test management (including Defect Management); as the primary repository for Test Artefacts and to manage • Requirements • Test Cases / Scripts • Test Plans • Defects
IBM Rational Requisite Professional (ReqPro)	Used for capture of business requirements; both functional and non-functional. Defect information can be fed back to ReqPro via automated *<Defect Management Tool>* workflow.
IBM Rational ClearCase	Software Configuration Management – code
IBM COGNOS	Data warehouse for *<name of project / program>* solution delivery metrics, including defect management
SERENA PVCS Version Manager	Version control – documents

In addition, a tool for managing CRs needs to be integrated with ReqPro or any SCM tools. The impact of CRs on requirements is then reflected in *<Defect Management Tool>* (Requirements tab) through its integration with ReqPro.

Deliverables

Identify the deliverables required for your Project / Program, for example:

The following items (as a minimum) will be delivered for *<name of project / program>*:

- Defect Management Plan (this document)
- Defect Management Process (MS PowerPoint and MS Visio files)
- Defect Manager Role Position Description
- *<Defect Management Tool>* customisation and agreed *<Defect Management Tool>* public defect metrics (per work stream, and program)
- Health Checks/Reviews and Internal Audits of Defect Management (per work stream, and program)
- Agreed schedule of Defect Triages (per work stream, and program)

Defect Reporting

Defect Metrics

All example metrics in this section are fictitious, and are used for illustrative purposes only. You will need to refine these once the *<name of project / program> <Defect Management Tool>* database is available for your set up and use.

This section focuses on defect metrics that can be produced from *<Defect Management Tool>* manually.

During test planning and test execution activity it is expected that defect metrics are produced on a regular basis. The exact frequency may vary and should be agreed with your Test Manager.

The level of reporting for defects may be for a specific work stream or for the entire Project / Program. This need will be accommodated through customisation of the specific *<Defect Management Tool>* database for *<name of project / program>*. Similarly, defect metrics for a specific test phase or application can also be produced.

In addition the Defect Manager will produce a weekly defect summary for all *<name of project / program>* work streams for Program Management (example only):

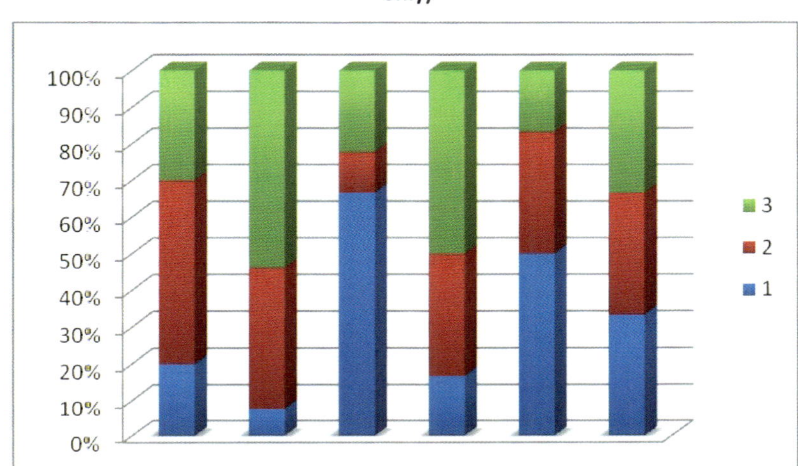

The categories within each work stream could represent any or all of the following possibilities:

- Total defects raised by priority

- Total defects raised by severity

- Outstanding defects raised by priority

- Outstanding defects raised by severity

- Total defects by test phase

- Oustanding defects by test phase

For metrics based on test phase it will be possible (for any given work stream) to compare the number of defects found by the technology test team and compare that with the number of defects detected by the business test team (for example in UAT).

It is important to be able to compare actual defect fix turnaround times against agreed defect Service Level Targets. A current *snapshot view* can be prepared for outstanding defects by comparing the defect age with the planned turnaround times for these defects based on Service Level Targets. Defect age for outstanding defects is based on the Date Detected and can easily be extracted from *<Defect Management Tool>* as a snapshot view each week or month and then broken down in terms of these severity and priority ratings, in order to highlight non-compliance with agreed defect fix turnaround Service Level Targets. This could be done at a work stream level or rolled up at a program level.   Refer *2.1.6 Service Level Targets* for assumed defect turnaround targets:

Table 15- Outstanding Defect Age by Severity and Priority (example only)

		1 day	2 days	5 days	1 month
Severity 1	Priority 1	1	1	0	1
Severity 1	Priority 2	0	1	1	0
Severity 1	Priority 3	0	0	0	0
Severity 1	Priority 4	0	0	0	0
Severity 2	Priority 1	0	0	2	0
Severity 2	Priority 2	0	0	2	2
Severity 2	Priority 3	0	0	0	0
Severity 2	Priority 4	0	0	0	0
Severity 3	Priority 1	2	0	3	0
Severity 3	Priority 2	0	0	0	0
Severity 3	Priority 3	4	3	7	5
Severity 3	Priority 4	2	1	6	5
Severity 4	Priority 1	0	0	0	0
Severity 4	Priority 2	0	0	0	0
Severity 4	Priority 3	2	3	4	5
Severity 4	Priority 4	0	1	6	4

In this example there is a persistent issue of failure to meet agreed defect turnaround Service Level Targets (red highlighted numbers).

Defect injection and closure rates can be highlighted at a workstream or program level, in order to highlight any worrying trends (example only):

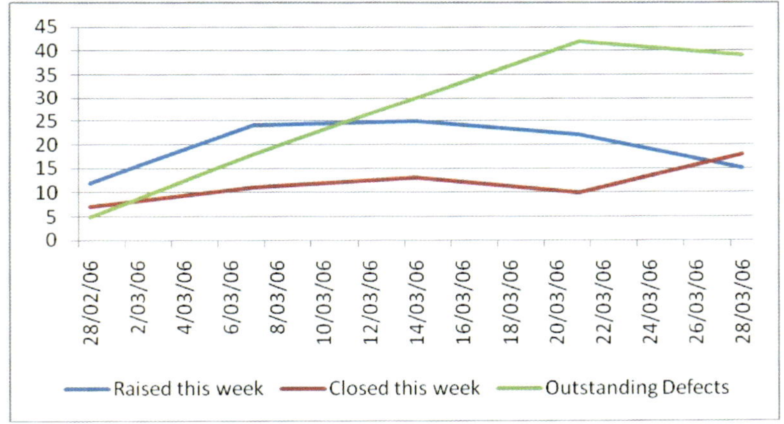

Figure 11- Defect Injection and Closure Rates (example only)

In this example the number of outstanding defects has increased rapidly, until the very last week. Perhaps additional development resources are now on hand, or perhaps the test team are nearing the end of test execution? On the other hand, if the worrying trend returns in subsequent weeks then this line graph could be highlighting a serious risk of project failure.

Other common defect metrics include defect density – usually this means the number of defects divided by the number of lines of code (or by the number of function points). However, variants (displayed by work stream) might include the ratio of defects to requirements.

For example, the following bar chart shows the number of requirements by workstream compared with the number of peer review defects ("proto-defects") and testing defects (example only):

Figure 12- Defect Density Based on Number of Requirements (example only)

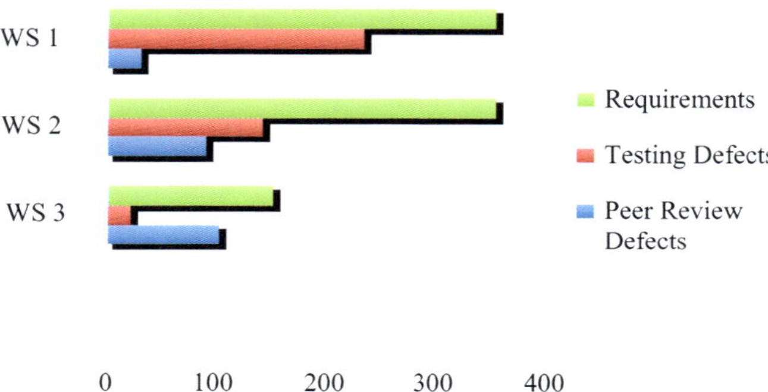

Defect metrics can also help in assessing Test Team efficiency as follows:

- Number of test cases run / number of defects found (this can be further broken down by test phase)

- Number of defects detected before production implementation / Number of defects detected during warranty

Table 16- Test Team Efficiency 1 (example only)

	WS 1	WS 2	WS 3	WS 4
Test Cases Run	400	350	120	276
Defects Found	40	77	21	46
Test Team Efficiency	10.00%	22.00%	17.50%	16.67%

Here a higher percentage is a better result, with testing finding more defects per test case on average. Often this metric is more useful in comparing test team efficiency for the same team across multiple test cycles or releases, rather than comparing test team efficiency across test teams.

Table 17- Test Team Efficiency 2 (example only)

	WS 1	WS 2	WS 3	WS 4
Warranty Defects	4	8	5	5
Testing Defects	40	77	21	46
Test Team Efficiency	10.00%	10.39%	23.81%	10.87%

Here a lower percentage is a better result, with a lower proportion of warranty defects to testing defects being preferable.

Also, as mentioned earlier, defect root cause analysis will be used by Test Managers and the Program Defect Manager (example only):

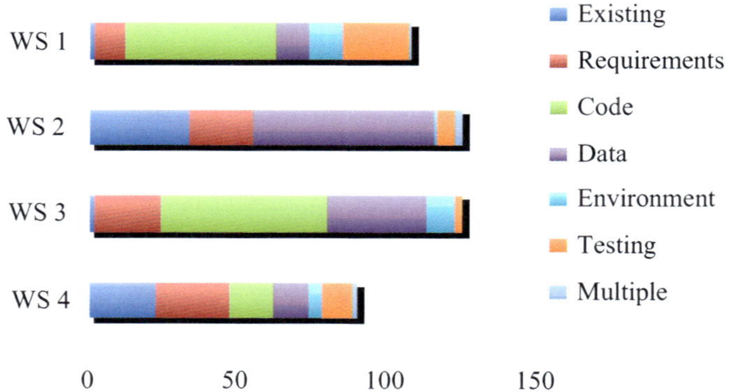

Note that peer review "proto-defects" will not be captured in *<Defect Management Tool>*. Hence, it will not be possible to use *<Defect Management Tool>* to compare the number of "proto-defects" detected during peer review activity with the number of defects detected during testing (for each work stream). However, it should still be possible obtain and compare "proto-defect" metrics from the Document Management process with *<Defect Management Tool>* defect metrics for the same work stream or product (for example) in order to assess any positive or negative trend in quality.

The details for each work stream in some of the example graphs above tell a different story and each set of raw defect metrics would need to be qualified through further analysis, consultation and interpretation before any firm conclusions could be made for each work stream.

Defect Management Audit

The purpose of a defect management audit is to assess compliance with, and assess the effectiveness of, the methods and processes outlined in this defect management plan. The key audience for a Defect Management Audit is the PMO.

An informal face to face survey of project managers, test managers and the release manager is recommended at the outset of any Defect Management audit in order to gauge whether or not compliance or effectiveness issues may exist.

For each work stream the Defect Manager should compile and compare the following metrics:

- Total defects by test phase

- Outstanding defects by age

- Defects by time taken to close (Closed defects only; duration in days = closure date – date raised).

- Current defect Service Level Targets

- Total requirements

- Total test cases

- Average defect triages per week

- Work stream test phase duration – planned versus actual

- Total number of issues (logged or not yet logged) related to defect management

Compliance

On a regular basis the Defect Manager should summarise any scheduled defect triages (by work stream).

On a regular basis the Defect Manager should spot check individual defect quality for each work stream.

For each work stream the Defect Manager should compare time taken for defects to be closed against defect Service Level Targets, and raise appropriate issues with the Technical Project Managers.

The Defect Manager should confirm with the Release Manager whether recommendations regarding the use of defect ids in check in comments and document histories are being followed, and whether these practices are adding any value.

The Defect Manager should summarise any Defect Management Audits conducted. A Monthly Defect Management Audit Report is recommended.

Effectiveness

One useful measure is to review the program issue log (open and closed) in order to assess whether or not any relate to defect management itself. If no issues have been raised then the defect management process itself has not been a problem (or at least was not reported as a problem).

From the metrics gathered, can any correlations be identified? Can any inferences be drawn? Can any cause and effect then be argued?   For example, just by focussing on defect triages, it may (or may not) be possible to establish whether:

- Teams with fewer defect triages per week took longer to close defects on average (and more frequently missed their defect fix Service Level Targets).

- Teams with fewer defect triages per week had older outstanding defects per week on average

- Teams with more defect triages per week had actual work stream duration closer to planned work stream duration.

- Teams with more defect triages per week had fewer cancelled defects as a percentage of their total defects raised

- Teams with more defect triages per week had fewer outstanding defects (as a percentage of their total defects raised) at production implementation (note that the only valid final statuses for a resolved defect are CANCELLED or CLOSED).

- Teams with more defect triages had fewer failed test deployments (e.g. requiring rework of release notes etc).

Another area to consider is the effectiveness of defect root cause analysis – did this activity help to identify any project issues that resulted in more defects? For example, did a particular team fail to follow the Peer Review process prior to releasing requirements for coding and test planning? Or did one test team regularly raise duplicate defects because they failed to check for duplicates before raising a defect (and the Test Manager failed to notice this problem)?

Testing Status Report

Work stream Test Manager weekly status reporting provides at a summary level, progress of testing.

Data is drawn from *<Defect Management Tool>*, which is the repository for test results and defects. Note that vendors may have their own defect management system – defects stored in *<Defect Management Tool>* will be in summary form and limited to those defects found during test which may impact the end-to-end solution (particularly design).

This is for limited distribution only, and communicates progress to the Program Test Manager during the test phase. Formal review (and communication of test progress to a greater audience) will be held weekly as part of work stream status meetings.

Refer Master Test Strategy for more complete details of the Testing Status Report.

Test Summary Reports

Reports are required at the end of each of the test phases, to summarise the current position of testing and allow stakeholders to determine if it is appropriate for the program to move to the next phase.

Test summary reports will be provided by the Work stream Test Managers for the Program Test Manager and work stream Project Manager, at the finalisation of each testing phase. These reports include the following defect details:

- Summary of outstanding defects including their severity and priority
- A detailed action plan for each outstanding defect

Refer Master Test Strategy for more complete details of the Testing Summary Report.

Defect Management Organisation

Roles and Responsibilities

Refer *<name of project / program>* RASCI maintained by PMO.

Refer Master Test Strategy for full details of testing responsibilities and *<name of company>* role mandates.

Refer Quality Management Plan, Quality Assurance Plan and Document Management Standard & Procedure for full details of Peer Review responsibilities.

Refer Release Management Plan for full details of Release Management responsibilities.

Table 18- Roles and Responsibilities – Defect Management

Role	Responsibility
Test Manager	The Test Manager shall: • Create Daily Status Reports during test execution (including defect metrics). • Create Test Summary Report (TSR) at the completion of test execution (including defect metrics). • Manage issues and changes in partnership with program management team. • Facilitate defect triages as appropriate, ensuring representation from all relevant teams. • Approve the Defect Management Plan • Review defects logged.

Role	Responsibility
Release Manager	• Participate in defect triages as required • Ensure that defect fixes are checked in (with defect ids in the check-in comments) prior to generating each build • Ensure that defect ids for fixed requirements defects are reflected in document histories • Review Release Notes (including defect details) from vendors / solution implementers and ensure requirements defects (against baselined documentation) also included.
Business Project Manager	• Participate in defect triages as required • Approve trivial change to be managed as a defect and not a CR
Technical Project Manager	• Participate in defect triages as required • Schedule time in the project plan for defect fixes • Review and approve contractual SLAs for defect fix turnaround times • Manage Solution Delivery staff (developers, business analysts etc) to ensure defects are fixed according to contractual SLAs.
Defect Manager	Reporting to the Test Manager, the Defect Manager shall: • Develop and manage the Defect Management Plan • Define *<Defect Management Tool>* customisation and automated workflow for defect management • Participate in tools integration between

Role	Responsibility
	<Defect Management Tool> / ReqPro / Clearquest / Clearcase etc as required • Review quality of defects across program and raise issues with PMs • Facilitate defect triages as required, ensuring representation from all relevant teams. • Coach Project Test Managers and Test Leads in defect triaging • Delegate defect triages to Test Managers as required, ensuring representation from all relevant teams. • Manage risks, issues and Change Requests coming out of defect management in partnership with program management team including QA Manager, Test Managers, Release Manager, Requirements Manager and Program Manager • Review contractual SLAs for defect fix turnaround times as required by TPMs • Define defect Service Level Targets (in this document) • Measure defect turnaround against Service Level Targets and raise concerns and issues with PMs • Produce program defect metrics • Conduct internal audits of the defect management process • Participate in external audits of the defect management process as required
Test Administration	Reporting to the *<name of project / program>* Test Manager, the Test Administration Coordinator (*<Defect Management Tool>*

Role	Responsibility
Coordinator - **(*<Defect Management Tool>* / Reporting)**	Reporting) shall, • Be the single point of contact for all Test Reporting activities • Monitor and drive the progress of Test Reporting activities • Create Weekly Status Reports during test planning and preparation • Create Daily Status Reports during test execution • Provide ad hoc reports from HP Quality Centre (*<Defect Management Tool>*) for management reporting • Provide the business specific reports as defined in the Defect Management Plan and Master Test Plan(s)
Test Analyst	Reporting to the Test Team Lead, the Test Analyst shall, • Log Defects when an expected result does not equal the actual result (includes setting initial defect severity and priority ratings). • Retest resolved Defects • Update Test Team Lead with progress and defect retest details
Business Analyst	• Review and Fix requirements defects • Update defects in *<Defect Management Tool>* • Participate in defect triages as required
Developer	• Review and Fix code defects • Update defects in *<Defect Management*

Role	Responsibility
	Tool> • Participate in defect triages as required
Test Environment Engineer	Working under the guidance of either the Test Manager or the Test Environment Manager, • Investigate and correct environment related Defects raised by testing. • Notify test analyst of test environment defect fixes
Test Environment Manager	Reporting to the Test Manager, the Test Environment Manager shall • Coordinate the timely response to environment related Defects raised by testing. • Ensure Configuration Management Processes are adhered to.
Project Test Manager	Reporting to the Program Test Manager, the project or work stream Test Manager shall • Create work stream Daily Status Reports during test execution (including defect metrics). • Create work stream Test Summary Report (TSR) at the completion of test execution at each test phase (including defect metrics) • Review defect quality of defects raised by work stream test team • Facilitate work stream defect triages as required, ensuring representation from all relevant teams. • Liaison with vendor development / support teams on defect resolution

Role	Responsibility
	• Manage the Defect process and resolution for their tests
Test Team Lead	Reporting to the work stream Test Manager, the Test Team Lead shall, • Ensure adherence to the Defect Management Plan and Master Test Plan by team members • Participate in defect triages as required

Impacts

Assumptions

Identify any assumptions you have in employing your defect management plan, for example:

Table 19- Assumptions

#	Assumption
1	There will be one *<Defect Management Tool>* database for the *<name of project / program>*
2	All *<name of project / program>* defects are to be raised in this *<Defect Management Tool>* database
3	In order to customise the database or add automated workflows to accommodate the approved defect management process, the Defect Manager will work directly with the *<name of project / program> <Defect Management Tool>* expert and the Test Manager.
4	Anyone who needs to log or update defects will be defined as a *<Defect Management Tool>* user with at least a Defect licence profile, and will have access to *<Defect Management Tool>* for *<name of project / program>*.
5	The Release Management Plan will be documented and implemented as briefly outlined in this Defect Management Plan.
6	Defect fix Service Level Targets documented in this Defect Management Plan are agreed for defect severity and priority and apply equally to code and requirements defects. These targets apply equally to all work streams in *<name of project / program>*. These are initial targets only and can be formally renegotiated and agreed at a later date.
7	There is no requirement to record Document Review Defect Sheets (or individual items on these sheets) as *<Defect Management Tool>* defects. This is technically easy to achieve, but not desirable.

Risks

Identify any risks you are likely to encounter in employing your defect management plan; for example:

The following risks have been identified that relate to defect management:

Table 20- Risks

No	Risk	Risk Likelihood	Risk Impact	Consequence	Risk Management Strategy	Responsibility
1	There are multiple defect management processes (and related *<Defect Management Tool>* processes) already at the *<name of company>*	Medium	Medium	May create confusion and resistance relating to this defect management plan.	Ensure key stakeholder review and sign-off of this defect management plan.	
2	Tools integration strategy in preliminary stages only	High	Medium	May hinder ability to "sell" defect management as part of this package	Obtaining more information	
3	Defect Service Level Targets / SLAs not met and/or unrealsitic	Medium	High	Not delivering on time Poor quality Software Impact to testing schedule	Application teams to follow defect turnaround/resolution as defined in the Defect Management Plan and in any contractual SLAs. Prioritise high complexity and priority areas to be tested as early as possible within the test phases.	
4	*<Defect Management Tool>* Remote Access is inadequate	Medium	High	Impact to project budget/ Impaired defect management / Impact to testing schedule	Remote Access is to be provided to test teams located outside the *<name of company>* Firewall with the appropriate quality of service. If remote access is unavailable, *<name of company>* to provide onsite facilities for *<name of project / program>* team.	
5	Inadequate *<Defect Management Tool>* availability or licencing	Medium	High	Impact to project budget/ Impaired defect management / Impact to build and testing schedules	Estimate and budget for sufficient concurrent licences (defect / full). Confirm process for deployment of *<Defect Management Tool>* thin client to SOE PCs.	

Issues

Identify any issues you have with defect management here, for example:

The following issues have been identified that relate to defect management:

All issues are managed in accordance with the <name of project / program> PMP and stored within the project / program's issues management tool. This section provides an overall assessment of the testing issues identified for this program. The table below is an extract and documents all known risks faced during the program testing and the approach to removing/mitigating these risks.

Table 21- Issues

#	Issue Description	Impact	Strategy to Manage the Issue	Owner
1)	There is no documented Release Management Plan ...	This creates uncertainty for defect management and change management	Documented an overview of Release Management from a defect management perspective in this document and documenting Release Management Plan.	
2)	*<name of project / program> <Defect Management Tool> project database not yet available*	Unable to configure *<Defect Management Tool>* to suit Defect Management Plan.	Working with Test Manager to resolve	

Dependencies

Identification of dependencies that impact on your defect management plan, for example:

Table 22- Dependencies

No.	Dependency Name	Dependency Description	Dependency Scope
1	*<Defect Management Tool>* availability	*<name of company>* to provide HP *<Defect Management Tool>* for all parties engaged as part of the *<name of project / program>* (including sufficient licences).	All releases
2	*<Defect Management Tool>* accessability	The strategy for *<name of project / program>* defect management relies upon *<name of company>* providing HP *<Defect Management Tool>* access to teams located outside the *<name of company>* Firewall	All releases
3	Quality Centre Reporting and defect workflow automation	Test Administration Coordinator required to implement HP Quality Centre customisation, defect workflow automation and reporting.	All releases
4	Release Management	For the successful management of defects, the Release Management Plan and Tools need to be in place and adhered to for both documentation and code.	All releases
5	Defect Management Support	Appropriate business and technical SMEs are available as support resources to review defects and partipate in defect triages as required.	All releases
6	IBM Rational RequisitePro and HP Quality Centre	Bi-directional synchronisation between ReqPro and *<Defect Management Tool>*. Defect details will be synchronised from *<Defect Management Tool>* to ReqPro.	All releases

Appendices

Appendix A - Definitions

Ensure that you provide a definitions of abbreviations and acronyms used in writing your defect management plan; for example:

Please ensure that all acronyms and specialist terminology in the document are covered off in this table. The following table provides definitions of acronyms and terms related to defect management on *<name of project / program>*:

Table 23- Definitions

Term	Definition
Anomaly	"Anything observed in the documentation or operation of software that deviates from expectations based on previously verified software products or reference documents" [Refer *IEEE*]
Baseline	A *Release Management* term relating to a set (or sub-set) of *configuration items* for a *product*, denoting a specific agreed state for that product (e.g. requirements baseline for first release of approved requirements, or code baseline for the first release of code to testing).
BRD	Business Requirements Definition
Change Item	A *Defect* or a *Change Request* that impacts one or more product *configuration items.* Never abbreviated to CI, to avoid confusion with *Configuration Item.*

Change Management	From a *Release Management* perspective, Change Management (also known as scope management, and Change Control) is a methodology comprised of a structured body of processes, procedures, working methods, practices and principles used to facilitate the efficient management of *product* related *Change Requests* (CRs) during testing, and into production.
Change Request	A request to change the scope of the baselined *product* requirements, subject to *Change Management* and *Release Management*.
CI	Configuration Item
Clearcase	IBM Rational Clearcase (an *SCM* tool)
Clearquest	IBM Rational Clearquest (*change management* tool, and proposed as process state engine)
Configuration Item	A software file or documentation file under *Release Management* for a *product*. Or a non-Product documentation file under *version control / document management.*
Configuration Repository	A Configuration Repository is a database containing a complete set of *Product* configuration items subject to *Release Management* and/or a complete set of non-Product configuration items (subject to *SCM* but not Release Management) that are nonetheless Project *configuration items.*
Controlled	A document is controlled when it is a project auditable document and is recorded in the *Document Register.*

CR	Refer *Change Request*
Defect	*"A product anomaly"* [Refer *IEEE*]. A defect is subject to *Defect Management* and *Release Management*. See also *proto-defect* and *informal defect*.
Defect Management	Defect management is a methodology comprised of a structured body of processes, procedures, working methods, practices and principles used to facilitate the efficient management of *defects* during testing.
Defect Triage	A defect triage is a best practice designed to ensure that the right *defects* are fixed, deployed to a test environment and re-tested first.
Document Register	A type of *configuration repository* subject to *version control* though not currently subject to formal *Release Management*. On *<name of project / program>* PVCS is used as a Document Register, where *configuration items* are classified as *Controlled*, *Records*, or *Registered*.
Dynamic Testing	Any testing that involves running test cases (manual or automated). Refer Master Test Strategy for more detail.
Feature	A term from Agile development meaning a functional component comprised of one or more *configuration items* for a *product*.
Foreign Defect Id	A custom field in the *<name of project / program>* *<Defect Management Tool>* database (refer *Working with the Vendor*)
Granularity	The level of detail at which a *defect* is raised

HPQC	Hewlett Packard <Defect Management Tool> - a test management and *defect management* tool.
IEEE	Institute of Electrical and Electronics Engineers. IEEE definition references in this document come from IEEE Standard 610.12-1990. *IEEE Standard Glossary for Software Engineering Terminology*.
Implementation Planning	Implementation Planning defines a plan of action to deploy and embed an initiative in the business. It documents how the implementation will be managed and the underlying logic behind this approach.
Informal defect	Defect noted by developer during unit testing but not formally raised in *<Defect Management Tool>* and not managed via the Defect Management process.
ITIL®	Information Technology Infrastructure Library – is the most widely accepted approach to IT Service Management in the world. ITIL provides a cohesive set of best practices, drawn from the public and private sectors internationally.
OLA	Operational Level Agreement
PVCS	SERENA PVCS Version Manager (an *SCM* tool used as a *version control* tool for *Document Register*). PVCS itself stands for Polytron Version Control System.
Peer Review	Documentation inspection technique as specified in *QA* Plan
PMBOK	Project Management Book of Knowledge

Project	"…a temporary endeavor undertaken to create a unique product, service or result." PMBOK
Product	This is specifically taken here to mean a software product, which comprises all of the *configuration items* (requirements and code) subject to *Release Management* in the product *configuration repository*.
Proto-defect	A type of *defect* detected early in the SDLC as a result of a Peer Review, captured via a Document Review Feedback Sheet, and managed via the *Document Management* process. Proto-defects may apply to both *product* and non-product *configuration items*. From the Greek for first, which is *protos*. Note that *proto-defect* is term of convenience for this Defect Management Plan only.
Quality Assurance	"A planned and systematic means for assuring that the outputs of a program, process, or other endeavor will fulfill the requirements of users, consumers or customers." (From the QA Plan)
Quality Management	"To ensure the program's deliverables meet the requirements and expectations set by the project's stakeholders." (From the QMP) "Management to apply Quality Assurance" (From the QA Plan)
QA	Quality Assurance.
QMP	Quality Management Plan.

Release	A release comprises one or more *defect* fixes and/or *CR* implementations, subject to *Release Management*.
Release Management	Release Management is a methodology comprised of a structured body of processes, procedures, working methods, practices and principles used for the management of software *releases* for one or more *products* during software testing and production implementations. Project Release management co-ordinates the activities of *Software Configuration Management, Project Change Management, Document Management* and *Defect Management.* Production Release Management co-ordinates the activities of *Software Configuration Management, Production Change Management, Incident Management, Problem Management, Service Transition* and *Implementation Planning.*
Release Note	A Release Note is produced for every *release* of software (and accompanying requirements documentation) into any test environment, and production. It details new / updated *Configuration Items* (including component version numbers), and *Change Items* (including defect ids and CR ids), together with any deployment / installation instructions.
Requirements Management	Requirements management is a methodology comprised of a structured body of processes, procedures, working methods, practices and principles used for the management of *product* requirements
SAD	Solution Architecture Design

SCM	"Software configuration management is a set of engineering procedures for tracking and documenting software throughout its life cycle, to ensure that all changes are recorded and the current state of software is known and reproducible."
SDLC	Software Development Lifecycle
Service Transition	An ITIL process, Service Transition is used to ensure that implementations of new or changed Technology services are transitioned into production in a controlled way so they can be operationally managed and supported. The focus is to minimise risk and maximise supportability of *<name of company>* Technology services.
Service Level Agreement	A contractually binding set of targets for defect turnaround times
Service Level Target	A consistent and measurable set of targets for defect turnaround times for all work streams on *<name of project / program>*, used to highlight issues or concerns with defect turnaround times.
SLA	Service Level Agreement
Software Configuration Management	Project deliverables are defined via Configuration Item Lists as Products and Non-Products. Product *change items* (*defects* and/or *CRs*) are bundled together with the associated product *configuration items* (requirements and code) into a *release* with accompanying mandatory *release notes* for deployment via the *build pipeline* to a test environment or production.

Sprint	A term from Agile development meaning a short-term development *release* comprised of a set of *features*.
Static Testing	Any activity that involves "testing" a product without running test cases – this can include Peer Review, test planning activities, or code inspections. Refer Master Test Strategy for more detail.
SoW	Statement Of Work
ReqPro	IBM Rational Requisite Professional – a *requirements management* tool
Version Control	A form of *SCM* whereby *configuration items* (*product* and/or non-product) are checked in and out of configuration repositories but are not necessarily subject to *Release Management*.

--- End of Defect Management Plan Template ---

Other e-Book Titles in the *Technology Template Series*

Project Management Deliverables

- ❖ Business Case - PM 001
- ❖ Project / Program Charter - PM 002
- ❖ PMP - Project / Program Management Plan - PM 003
- ❖ Steering Committee Charter - PM 004
- ❖ Steering Committee Dashboard - PM 005
- ❖ Communications Plan - PM 006
- ❖ Risks Register - PM 007
- ❖ Issues Register - PM 008
- ❖ Deliverables Register - PM 009
- ❖ Dependencies Register - PM 010
- ❖ Dependency Agreement - PM 011
- ❖ Change Management Strategy - PM 012
- ❖ Change Management Plan - PM 013
- ❖ Implementation Plan - PM 014
- ❖ Training Plan - PM 015
- ❖ Post Implementation Review - PM 016
- ❖ Schedule - PM 017
- ❖ RACI - PM 018
- ❖ Resource Management Plan - PM 019
- ❖ Meetings - Agenda - PM 020
- ❖ Meetings - Minutes - PM 021
- ❖ Review Feedback Form - PM 022

Technology Deliverables

- ❖ Business Requirements Definition - TD 001
- ❖ Business Detailed Requirements - TD 002

❖ Solution Architecture Design - TD 003
❖ Software Requirements Specifications - TD 004
❖ Detailed Design Document - TD 005
❖ Technical Specifications - TD 006
❖ Support Plan - TD 007
❖ Implementation / Deployment Plan - TD 008
❖ Operational Support Readiness Plan - TD 009
❖ Readiness Cutover Report - TD 010

Test Management Deliverables

❖ Test Strategy - TM 001
❖ Master Test Plan - TM 002
❖ Schedule - TM 003
❖ Resource Management Plan - TM 004
❖ Roles & Responsibilities Definition - TM 005
❖ RACI - TM 006
❖ Defect Management Plan - TM 007
❖ Release Management Plan - TM 008
❖ Requirements Traceability Matrix - TM 009
❖ Test Environment Plan - TM 010
❖ IPTA - TM 011
❖ Data Management Plan - TM 012
❖ Test Estimation Model - TM 013
❖ Detailed Test Plan - TM 014
❖ Performance Testing Plan - TM 014
❖ Test Readiness Review - TM 015
❖ Test Summary Report - TM 016
❖ Test Status Report - TM 017
❖ Test Case Review Log - TM 018
❖ Test Coverage Matrix - TM 019

About The Author

Ronald is a highly experienced and innovative IT professional with extensive Project Management, Testing and QA knowledge derived from over 30 plus years in providing evaluation and establishment of effective testing and risk management strategies. Ronald's core IT competencies are in the areas of Test Strategies, Methodologies, Systems and Integration Testing, Project/Program Management/Directorship, Service Delivery Management, Quality Assurance, Change and Problem Management. His extensive experience is underpinned by a sound background in IT and experience with a variety of technologies i.e. legacy systems, client/server, Web, E-Commerce, and Networks.

Ronald has successfully delivered to multiple programs/projects, constantly achieving effective business solutions on projects and programs up to $1+billion, providing a depth and real value to organisations. He has managed external Vendors on multiple projects and has off shore experience. Ronald thrives on being an enabler for the business through the stabilisation of IT and standardisation of processes. He delivers a comprehensive roadmap for delivery to the business, whilst delivering against milestones. These incorporate Business Continuity Planning, Disaster Recovery, Risk Analysis, Business Process Reengineering, IT Policies and Work Flow Analysis.

--- End of Document ---

Printed by Amazon Italia Logistica S.r.l.
Torrazza Piemonte (TO), Italy